New Product Development for Small and Medium Enterprises

A practical application of new product development concepts to empower small and medium sized enterprises.

Dr. Thomas V Edwards Jr

New Product Development for Small and Medium Enterprises

Dedication

This book is dedicated to my wife Deborah. Words are a poor means of expressing the debt of gratitude I will forever owe to Deborah for her love and support. Not only would my doctoral journey at the heart of this book have been impossible without her patience, support, and encouragement but her dedication to her own art has been a lifetime inspiration to me.

Acknowledgements

This work would not have been possible without the knowledgeable guidance and patience of several key individuals. First is my wife Deborah who did the first edit of the manuscript and helped me strike a better balance between the rigorous language of a scholar and the ability to communicate with everyone else in the world.

I would like to acknowledge and thank every engineer, scientist, and manager that I interacted with in my industrial new product development career. My apologies to everyone who has encouraged me over the years who I have not recognized by name. The oversight is due to my limited memory and not the importance of your contribution or the gratitude that I owe you.

Lastly, I will forever be indebted to my parents for many things including a deep appreciation for the value of the education they could not access in depression-era Philadelphia and Appalachia.

Table of Contents

About the Author

Dr. Thomas V, Edwards, Jr., DPS approaches the challenge of technological innovation in organizations from the three perspectives of research, teaching, and practice. Dr. Edwards earned a doctorate in management from Pace University, an MBA from LaSalle University, a master of mechanical engineering from Villanova University and a bachelor of science in mechanical engineering from Drexel University. Dr. Edwards currently is the chair of the Engineering, Technology, and Management Department at the Temple University College of Engineering in Philadelphia. Dr. Edwards's long industrial career included positions such as project engineer, chief engineer, engineering manager and general manager at companies in the aerospace and capital goods industries. Dr. Edwards' professional experience, doctoral research and teaching efforts all focus on the challenge of making our organizations more innovative as a path to sustainable competitive advantage.

Introduction to SME Innovation

I began my engineering career with a passion for developing new things. In my first engineering job I quickly discovered that engineering is the art, science, and politics of bringing things into the world that never before existed! I honed this insight into a skillset on subsequent development projects including the Thiokol Star 6B rocket motor and the DARPA Small Caliber Smart Munition. As my career progressed, I took a general manager's job with profit and loss responsibility for a business that designed, fabricated, and commissioned custom capital equipment systems for industrial customers. Every system we shipped was one-of-a-kind. Here I learned first-hand the business challenges of depending entirely upon brand new engineered-to-order products. Normally valuable productivity tools like six-sigma are not very useful when every shipment is essentially a brand-new product. The company needed to be good at anticipating errors rather than detecting and correcting them. Processes such as design reviews and the application of past lessons-learned to new designs become more important than statistical process controls. After ten years of P&L leadership of my division I was asked to take on the challenge of developing a new product development operation at the corporate level. I immediately dove into everything I could read about innovation and new product development, including Cooper, Christensen, and every Harvard Business Review article that I could find. But I began to question what I was reading. How do I know this is valid? How do I separate the truly applicable from the merely anecdotal? What is applicable in one context but not another? It turns out that these are risky questions to ask as I eventually devoted the next seven years to the pursuit of doctoral studies in innovation while simultaneously creating a new product development department at a NYSE listed company. These parallel efforts empowered me to connect the spheres of research theory and industry practice. I have created this book to help the management of Small and Medium Enterprises to understand and meet the challenge of new product development. The research underlying this book was first presented at the American Society for Engineering Management 2017 International Annual Conference. This book expands on the research to provide a practical guide for SME leadership to use new product development as a tool to improve the competitiveness of their company. I do hope that you find these insights both interesting and useful. Please contact me (Tom Edwards) at tve@temple.edu with any questions regarding this book. All the best to you in your pursuit of sustained competitiveness through new product development!

Chapter 1 The Nature of the Challenge of SME Innovation

Introduction

Managers of Small and Medium Enterprises (SMEs) are often faced with the challenge of developing and introducing innovative new products and services to increase the competitiveness of their firms. There is an abundance of literature on innovation but unfortunately it has been developed primarily by researching large firms. The results of these studies must be applied cautiously to SMEs. There is some research literature that specifically addresses SME innovation but mostly for the purpose of identifying the barriers faced by these firms. There is sparse guidance to the SME management on how to overcome these challenges. I used the case study method of investigating new theory to examine potential solutions to the challenges of innovation by SMEs. Management research is broadly divided into two approaches. Quantitative research relies on the statistical analysis of data to test hypotheses based on existing theory. Qualitative research is used to develop new theory that can be subsequently tested. Since there is limited theory as to how SMEs should innovate, I used a qualitative approach, the case study method, to develop new theory. The new theory that emerged from this effort provides SME managers with guidance to help meet the innovation challenge.

Literature Review

Any research project starts with a review of the existing theoretical literature to establish the exact state of our current knowledge. During my doctoral studies of innovation, I found that much of the innovation research focused on larger firms[1]. However, SMEs are an important part of the overall economy that may require a different innovation management approach than larger firms. The first challenge is to define an SME. The United States Government defines an SME as a firm with fewer than 500 employees. This is quite a large range and it is not reasonable to assume that management approaches that work with a five-person firm will work with a 500-person firm. If we assume that SMEs must be at least a 100-person firm to have the resources to innovate and the business complexity to require specific management approaches, then this would represent over 82,000 firms in the U.S. according to census data.

SME firms also make a significant contribution to the total innovative output of the economy. SME firms spent 24% of all the R&D spending in 2005 in accordance with a 2009 study by researchers Van de Vrande, De Jong, Vanhaverbeke, and De Rochemont. This is a significant increase from the 4% in 1981 determined by the National Science Foundation. The Van de Vrande et al study went on to enumerate the unique challenges faced by SME firms when they innovate, including a lack of financial resources, the difficulty in recruiting specialized workers, a small innovation portfolio making it difficult to diversify project risk, and the need to draw heavily on their external network to find the missing innovation resources. Scholars Scozzi, Garavelli, and Crowston in a 2005 study add to the discussion by noting an inadequacy of management and marketing skills that that often result in an SME being unable to exploit new product developments.

Rosenbusch, Brinckmann, and Bausch in 2011 aggregated 42 empirical studies of over 21,000 firms and found that innovation has a positive effect on the financial performance of SME firms and that managing the process of innovation carefully is important to SME firms. They concluded that the dedication of substantial resources to the innovation efforts was necessary but not sufficient for the success of SME firms, as SME firms need to pay close attention to the process of how they manage innovation. The authors found that the unique challenges faced by an SME firm attempting to innovate include a lack of resources, potential existential risks if an innovation project should fail, and a lack of organizational capabilities to execute the innovation successfully.

Scozzi et al (2005) added to our understanding of the drawbacks facing SME firms as they attempt to innovate: noting that they often do not develop an innovation strategy, do not methodically attempt to control and monitor the innovation process, do not attempt to capture the knowledge acquired during development efforts, and SME firms were unable to identify problems that occurred during innovation developments. Despite these problems, the same authors note that some SME firms are successful innovators taking advantage of their small size and nimble decision-making processes to innovate successfully to serve attractive niches with innovative products in a way that larger firms cannot.

Huang, Soutar, and Brown (2002) investigated if the SME firms follow the 13 new product development steps argued for by Cooper and Kleinschmidt (1987). They found that SME firms were good at technology related activities such as product development, in-house product testing, and preliminary technical analysis. They also found that SME firms were not effective at marketing related activities such as market studies, market testing, and preliminary marketing analysis. Nevertheless, they concluded that although SME firms are not generally effective at the marketing related activities it is the marketing related activities that result in successful new product commercialization efforts.

In summary, innovation can be quite valuable to SME firms but they have specific challenges to overcome as summarized in Exhibit 1. There is insufficient understanding of how to overcome these unique barriers faced by SME firms attempting to innovate and a better understanding is important given the large number of SME firms and the fact that they spend almost one-quarter of the total R&D spending in the United States (Van de Vrande et al, 2009). SME firms present unique advantages and disadvantages and the management of their innovation efforts need to be better understood. The next chapter will examine the case study approach that I used to assist SME managers to become more effective innovators.

Exhibit 1. Challenges facing SME firm innovation efforts

Challenge	Reference
A lack of financial resources	Van de Vrande et all (2009), Rosenbusch, Brinckmann, and Bausch (2011)
A lack of organizational capabilities to execute the innovation successfully	Van de Vrande et all (2009), Rosenbusch, Brinckmann, and Bausch (2011)
Small innovation portfolios make it difficult to diversify project risk	Van de Vrande et all (2009), Rosenbusch, Brinckmann, and Bausch (2011)
SME firms need to pay close attention to the process of how they manage innovation • Strategy • Controlling and monitoring • Identify problems during innovation projects	Rosenbusch, Brinckmann, and Bausch (2011), Scozzi et al (2005)
SME firms are not effective at marketing related activities such as market studies, market testing, and preliminary marketing analysis.	Scozzi, Garavelli, and Crowston (2005), Huang, Soutar, and Brown (2002)

There is one exception to my comment that research has not determined how SME firms can manage the challenge of innovation. The work of Gomes-Casseres (1997) argues that the key difference is not the absolute size of an SME firm but rather its relative size compared to its market rivals. While some SME firms are tiny compared to their larger rivals there are some that are relatively large players in their specific niche. Despite their small size, they are dominant players in these niche markets. The author suggests that this deep-niche strategy is characterized by three elements: SME firms that are large relative to their direct competitors, SME firms that are technological leaders in their industry niche, and a focus on a limited group of industrial buyers. The focus on a limited group of industrial buyers precludes the need of the SME firm to invest in an extensive distribution network or expensive advertising. Their sales strategy consists of maintaining a technological leadership position and in maintaining key relationships with a handful of multinational buyers. Any firm employing the deep-niche strategy will rely on in-house capabilities to compete in a narrow market rather than engage in alliances which will erode their control over key sources and technologies. This is sound strategic advice if your firm satisfies the three constraints in the Gomes-Casseres model. If you find yourself in such a market your fundamental strategy will be to establish, or maintain, technological superiority.

References

Cooper, R. G., & Kleinschmidt, E. J. (1987). New products: what separates winners from losers? *Journal of product innovation management*, *4*(3), 169-184.

Gomes-Casseres, B. (1997). Alliance strategies of small firms. *Small Business Economics*, *9*(1), 33-44.

Huang, X., Soutar, G. N., & Brown, A. (2002). New product development processes in small and medium-sized enterprises: some Australian evidence. *Journal of Small Business Management*, *40*(1), 27-42.

National Science Foundation, 2006. Science Resource Studies, Survey of Industrial Research Development

Rosenbusch, N., Brinckmann, J., & Bausch, A. (2011). Is innovation always beneficial? A meta-analysis of the relationship between innovation and performance in SMEs. *Journal of business Venturing*, *26*(4), 441-457.

Scozzi, B., Garavelli, C., & Crowston, K. (2005). Methods for modeling and supporting innovation processes in SMEs. *European Journal of Innovation Management*, *8*(1), 120-137.

Van de Vrande, V., De Jong, J. P., Vanhaverbeke, W., & De Rochemont, M. (2009). Open innovation in SMEs: Trends, motives and management challenges. *Technovation*, *29*(6), 423-437.

Chapter 2 - Using Case Study Research to Investigate SME Innovation

Methodology

In a previous chapter, we established that Small and Medium Enterprise (SME) firms face unique challenges when attempting to innovate. We also found that the existing research literature does not recommend how SME managers can overcome these barriers. I set out to perform a study to provide SME managers with these recommendations. Since there are no existing theories of SME innovation management there is nothing to statistically test. Rather I designed a case study approach in accordance with Eisenhardt (1989) and Glaser and Strauss (1967) to develop new theoretical constructs of how SME firms can manage innovation. Eisenhardt argues that "building theories from case study research is most appropriate in the early stages of research on a topic" and she encourages the examination of emerging concepts by comparing them to the enfolding literature and determining: "what is it similar to, what does it contradicts, and why".

The author was a direct participant in the focal case study with deep first-hand knowledge of a specific episode of managing innovation at an SME firm. This unique access to data has been called "retrospective action science" by Gummersson (1999), who argued that this access and pre-understanding should be made use of by the scientific community claiming there is a "wealth of information stored in the minds of people who have lived through important and often dramatic changes. They did not at that time see themselves as researchers but afterwards started to reflect on what they have been through, asking themselves the question, could this be of general interest to the disciplines of management and economics". The author of this paper was assigned by the CEO of an SME firm to lead the development of a new product development department at the corporate level where one had not previously existed, therefore satisfying the requirements of unique access and pre-understanding.

The details of the focal case study are put into the context of the organizational situation so that the reader can understand how the situation of interest emerged as recommended by Klein and Myers (1999). My study followed the recommendations of Rosseau and Fried (2001) and generated new theory by linking case study observations to the theoretical constructs developed in the literature review. This approach also encompasses the previously discussed call by Eisenhardt (1989) to compare concepts emerging from the case study to the extant enfolding literature.

This case study is set within the theoretical framework of creating an innovative capability at an SME firm. The next section will present the details of the focal case study that resulted from the author's firsthand experience in managing the SME innovation program. The time frame of this case study starts with the initial kickoff of the program and concludes approximately six years later when the author was no longer involved with the program.

The Case

The focal company of this case study (hereinafter referred to as the company) was a successful manufacturing firm that had delivered solid financial performance for over 40 years by supplying a variety of products in the handling of specialty fluids such as corrosive or high temperature gases and liquids.

The focal company is in the center range of the U.S. Government definition of an SME as it is less than 500 employees with annual revenues less than $100M USD.

Despite the long-term financial success of the company, it enjoyed a limited history of innovation. At one point in its history the company had supported the internal development of a specialty fiberglass product that eventually grew into one of the most successful business units. More recently the company had successfully utilized a replacement raw material for that same specialty fiberglass product in response to more stringent environmental regulations. Another business unit had also successfully developed a biological based product to augment its chemical-based emissions control products. These were the sum total of the innovations introduced by the company despite its excellent history of operational excellence in closing sales and delivering products.

The CEO decided that the company needed to add the ability to develop innovative new products on a regular basis to supplement the aforementioned operational excellence. The first step was to form an executive committee to review an employee suggestion program. The employee suggestion program was thoroughly designed with a stage-gate evaluation of ideas and recognition, and cash awards for employees who submitted ideas that move through the stages and gates to eventual commercialization. This executive steering committee was composed of high stature individuals within the company, including three business unit general managers, a senior financial executive, a senior procurement executive, and a university researcher with long-time service on the company's board of directors. As the ideas made their way through the various stages the projects were evaluated on the basis of the development budget, the development schedule, and a cash flow model based on the envisioned commercialization. These items began as an informal approximation at the first gate and became subsequently more refined as the idea moved through the stages.

The author of this book was asked to lead these innovation efforts. Although assigning a senior executive to focus 100% of their time on new product development may seem unusual for an SME firm, the budget to support both myself and the development was less than 1% of the annual revenues of the company.

The employee suggestion program was kicked off aggressively with the CEO making his support of the project quite evident. The CEO traveled to the different business unit locations to announce the employee suggestion program. This initial effort developed a reasonable number of suggestions but an unanticipated result was that most of the suggestions dealt with operational efficiency issues that were not the objective of the employee suggestion program. Consequently, there was no mechanism to handle these suggestions as the steering committee was only empowered to invest in new product ideas. An unfortunate side effect of this is that employees who submitted suggestions to improve the operational efficiency of the company were disappointed when their suggestions were not acted on. This reduced their enthusiasm to make future suggestions. This suggests that if you ask employees for suggestions, you must be prepared to respond to a wide variety of suggestions.

The new product development process was initially envisioned as being entirely driven by employee suggestions but the effort evolved along a different path. Exhibit 2 summarizes the source of new product suggestions after approximately one year of operating the suggestion program.

Exhibit 2. Source of employee suggestion ideas and status after one year

Idea originator	Ideas judged worthy of eventual development	Ideas actively developed	Ideas launched
General manager/vice president	7	2	
Assistant general manager	1	1	1
Sales manager	3		
Chief engineer	1	1	
Regional sales manager	1		
Engineer	2	1	
Applications engineer	1		
Technician	1		
TOTAL	***17***	***5***	***1***

It is interesting to note that 66% of the ideas that passed the steering committee evaluation were submitted by the top leadership of specific business units (general manager, sales manager, or chief engineer). However, a quantitative count of suggestions does not account for the quality of the suggested ideas. Over the course of the program, one of the top three ideas was suggested at the top of a business unit (the general manager) while another of the top ideas was suggested from the working-level of the business unit (a sales engineer). These results lead to the argument that although an employee suggestion program should encourage suggestions from all levels of the organization, the top leadership may account for a high percentage of quality ideas since they are very close to customers and technologies.

After the first year of operation the original surge of suggestions began to subside and it became obvious that other avenues of ideation were required. We added three additional ideation techniques: a two-phase facilitated brainstorm process at each business unit, deep-dives into strategically important areas, and scouting of university technology development and licensing organizations. Additionally, the stage-gate process was redesigned to capture ideas for future revaluation if they were not acted upon at the time of suggestion. And new product development roadmaps were developed for each business unit to understand how one innovation could flow into another

The two-phase facilitated brainstorm process started with the sales department in each business unit. The idea was to develop a list of new products or services that could grow the business unit's revenue and profit. This list was developed by working with the individuals who were face-to-face with customers on a daily basis and were measured on their ability to generate revenue. The second phase was to take this list to the engineering and production departments of the business unit and brainstorm how to develop and deliver these innovations that the sales department believed were valuable.

An example of a deep-dive into a strategically important area was the suggestion that the company could produce a system for the desalination of brackish groundwater by the use of components from various business units within the company. The deep-dive consisted of developing a preliminary conceptual skid design and discussing it with various players in the industry such as desalination membrane suppliers, engineering firms that serve this market, and potential end-users such as water utilities and industrial users of water. Although the eventual conclusion was that the entire skid did not make sense from a business perspective, a project was spun off to develop a new-to-the-world corrosion proof component for the very front-end of any system dealing with brackish groundwater. This product is one of the significant success stories from this effort and it came from a deep-dive into a strategically important area.

Exhibit 3 is a summary of the total number of ideas worked on as part of the program. The ideas are sorted by the source of the idea (employee suggestions, scouting, deep-dives, brainstorming) and whether are not they were ultimately successful. Success is defined as resulting in a product that could increase the revenue or profit of the company. Some of the projects yielded valuable information but not a revenue generating product. These projects were not counted as a success. Two examples from the very early days of the program illustrate this concept. One idea that was investigated and tested was the use of a super-austenitic stainless steel as a replacement for Hastelloy at about half the material cost. This project concluded that we could make the substitution and thus either harvest more profit from a project or bid more aggressively and win more projects. This was counted as a success. Another example was an idea to determine if stainless steel ductwork in a key market could be replaced by carbon steel. This involved field testing of corrosion coupons in the environment of interest and subsequent evaluation of the coupons. This project concluded that carbon steel was not a good material for this application. The results of this project were valuable in terms of avoiding an expensive field failure and also providing the sales personnel with data to educate customers as to why the more expensive stainless steel was necessary. However, this project was still not counted as a success since it did not result in an increase in revenue or profit. The ideas are also evaluated for whether their contribution was incremental or disruptive. For the purposes of this paper incremental is considered an improvement to an existing product or an additional product in an existing product line. A disruptive innovation is defined as one that results in a unique value proposition to the end-user that competitors will have difficulty emulating. Ideas rated as inconclusive were still being pursued at the end of the case study timeframe and it was not yet obvious if they should be rated as a success or failure.

Exhibit 3. Overall results of the 6-year innovation program

Idea Source	Impact	Ideas acted on	Success	Fail	No conclusion
Employee suggests	Increment	13	10	2	1
	Disruptive	7	3	2	2
Tech scouting	Increment	0	0	0	0
	Disruptive	5	1	4	0
Deep dive	Increment	0	0	0	0
	Disruptive	2	1	1	0
2-phase brain-storm	Increment	4	2	0	2
	Disruptive	2	0	0	2
TOTAL		**33**	**17**	9	7

Although a higher success rate of the incremental innovations suggested by Exhibit 3 is not unexpected, the relatively high percentage of disruptive innovations that were successful is noteworthy. The disruptive innovations account for 29% of the successful innovations driven by this program. Furthermore, of the five disruptive success stories, three had been awarded utility patents by the USPTO and an additional project was still making its way through the patent process further supporting the conclusion they were sufficiently novel and non-obvious to be disruptive. Furthermore, one of the five disruptive products was associated with 24% of the total revenue of the relevant business unit. This specific innovation is an example of an architectural innovation (Henderson and

Clark, 1990) which will be addressed in detail in a future paper.

Subsequent chapters in this book will dive deeper into the case story outcomes and compare them to existing theory in the research literature. The purpose will be to develop new theoretical constructs for the management of SME firm innovation.

References

Eisenhardt, K. M. (1989). Building theories from case study research. *Academy of management review*, *14*(4), 532-550.

Glaser, B., & Strauss, A. (1967). The discovery of grounded theory. 1967. *Weidenfield & Nicolson, London*, 1-19.

Gummesson, E. (1999). *Qualitative methods in management research*. Sage Publications, Incorporated.

Henderson, R. M., & Clark, K. B. (1990). Architectural innovation: The reconfiguration of existing product technologies and the failure of established firms. *Administrative science quarterly*, 9-30.

Klein, H. K., & Myers, M. D. (1999). A set of principles for conducting and evaluating interpretive field studies in information systems. *MIS quarterly*, 67-93.

Rousseau, D. M., & Fried, Y. (2001). Location, location, location: Contextualizing organizational research. *Journal of organizational behavior*, *22*(1), 1-13.

Chapter 3 – Results of Case Study Research

Discussions and Conclusions

In Chapter 1 – *The Nature of the Challenge*, we summarized what is currently proposed about how SME firms need to innovate. The following table compares these conclusions in the existing literature with the results of the case study I described in Chapter 2 entitled *Using Case Study Research to Investigate SME Innovation.*

Exhibit 4: Evaluating claims in the literature to results from the case study.

Claims in the literature	Are the claims supported	Case study concepts
SME firms lack the financial resources to innovate	No	The results of this case study do not support the argument that a lack of financial resources should inhibit innovation by SME firms. The firm in this case study enjoyed considerable success in the development of innovative products without a massive investment relative to the size of the firm.
SME firms lack the organizational capabilities to successfully	No	The results of this case study do not support this argument. The organizational capabilities

Claims in the literature	Are the claims supported	Case study concepts
innovate.		required for innovation can be marshaled by an SME firm. If the firm can engineer and produce a specific product then it appears the firm can support innovations based on that product. Some external resources may be required but these can be reasonably integrated into the efforts of the SME firm.
A small innovation portfolio makes it difficult to diversify project risk	No	Again, the results of this case study do not support this argument. The focal SME firm in this case developed a sufficient portfolio of projects and they enjoyed considerable success despite the inevitable failures that must be anticipated in any innovative effort.
SME firms need to pay close attention to the process of how they manage innovation • Strategy • Controlling and monitoring Identify problems during innovation projects	Yes	The results of this case study support this argument. The focal firm in this case had an innovation strategy crafted for the specific circumstances of each business unit. The results of innovative projects were monitored and corrective action taken when necessary. Additionally, problems were identified in the innovation process itself

Claims in the literature	Are the claims supported	Case study concepts
		and corrected. This resulted in a modification to the initial stage-gate process as well as the development of a product launch methodology based on lessons learned.
SMEs can depend on a deep-niche strategy rather than challenge larger rivals. • SME firms that are large relative to their direct competitors, • SME firms that are technological leaders in their industry niche, SME firms focus on a limited group of industrial buyers	No	The results of this case study do not support that the only option open to SME firms is to adopt the deep-niche strategy and avoid challenging larger rivals. The focal SME in this case study successfully challenged several larger rivals by relying on the architectural model of innovation. A future chapter will further explore the application of architectural innovation.
SME firms were not effective at marketing related activities such as market studies, market testing, and preliminary marketing analysis.	Partial	The results of this case study partially support this argument. The lack of effective marketing related activities was certainly evident at the beginning of this case study. However, the focal SME in this case study was able to overcome this limitation and become effective at the marketing

Claims in the literature	Are the claims supported	Case study concepts
		related activities. This suggests that perhaps a lack of skill in these marketing related activities represents the lack of opportunity for learning by SME firms rather than an inherent lack of capability.

The results of this case study suggest a previously unpublished theoretical model for SME firm innovation consists of seven elements.

1. SME firms are fully capable of innovating at an investment level that is reasonable compared to the revenue of the firm.
2. Organizational capabilities may be less of an issue then the ability to focus these capabilities through an established new product development office to support the innovation.
3. SME firms can develop a sufficient portfolio of projects to innovate despite the inevitable failures that must be anticipated in any innovation effort. SME innovation calls for a balance of disruptive and incremental innovations.
4. SME firms must pay close attention to the process of innovation and how they manage the process.
5. Architectural innovation is a potential strategy for SME firms to challenge larger rivals.

6. The required marketing capabilities can be learned by SME firms.
7. An SME firm that is not the technological leader in a deep-niche market needs to first focus on establishing technological superiority over its rivals.

One limitation of this case study is related to the very definition of an SME. A small business is defined by the United States Government as a business with less than 500 employees. As noted previously this is a very large range and what is true for a 500-employee company may not be true for a firm with only five employees. It should be noted that the focal SME in this case is in the larger half of this range. However, several of the business units that successfully innovated were in the range of 25 to 50 employees. This extends the conclusions of this case study research into the smaller size of SME firms. The results of this case study should be applied cautiously to very small organizations.

Chapter 4 – Application of the Lean Startup Method to SME Innovation

INTRODUCTION

This chapter is based on a very successful workshop that I facilitated at a Small and Medium Enterprise (SME) firm to guide a new product launch. The workshop is based on the Steve Blank Lean Start-Up methodology with modifications to meet the specific needs of an SME manufacturing firm serving a business-to-business market niche.

Unfortunately, there is no "standard formula" for marketing at SME firms. There is an abundance of research into large firms that market consumer goods (the entire field of "consumer behavior"). There is also a significant amount of work on marketing by entrepreneurial start-up firms exemplified by the work of Steve Blank. However, there is little useful work on SME firms that provide engineered products in a business-to-business market.

I am convinced of the efficacy of the Steve Blank start-up approach based on the results I have witnessed as a grant reviewer for the National Science Foundation. I argue that it will be easier to adapt the Steve Blank approach to SME needs than it will be to adopt the large company consumer goods approach. There are similarities between a startup and an SME such as budget limits and addressable markets that tend to be narrow niche markets. I recommend that SME firms modify the Steve Blank approach to their specific needs.

THE SITUATION

The company that is the focus of this chapter (hereafter referred to as the focal company) was an SME firm that manufactured sophisticated electromechanical subsystems for use in their customer's factory automation systems. The focal company was evaluating the option of integrating forward and manufacturing their own factory automation system making use of the key subsystem that they provided. After navigating the complex politics of potential competition with major customers, the focal company decided to proceed with the project. The author then designed and facilitated the workshop to direct the development of the new product.

DESCRIPTION OF THE WORKSHOP

Company leadership first nominated who would participate. The sales and marketing functions were more heavily represented since the primary question was "what to sell" rather than "how to build it". We included "sanity checks" by the engineering and production departments as required. Participants were assigned pre-work so they were knowledgeable about the basic concepts prior to each workshop meeting. The weekly workshop meetings focused on applying the week's concepts to the marketing of the new factory automation product. One of the major tenants of the Steve Blank approach is to get "out of the building" and test assumptions by talking to customers, suppliers, etc. In every workshop meeting the participants suggested out-of-the-building actions to test the conclusions of each week's workshop. Company leadership reviewed the suggested out-of-the-building actions to prevent interference with on-going sales or project work. The results of the out-of-the-building testing was reviewed at the start of the next workshop meeting. The final deliverable was a Business Model Canvas (Figure 1) that captured an appropriate business model for the focal SME to address the factory automation market. A secondary deliverable was the development of a process for other business development projects.

THE WORKSHOP

The workshop proceeded as shown in the table below. All participants were assigned pre-work so that they were familiar with the concepts of each workshop. Each workshop lasted one or two hours. The post-work actions were assigned at the workshop meeting. Results were to be distributed prior to the next workshop meeting

Exhibit 5 – Outline of lean startup workshop to SME new product development

Workshop topic	Post-work	Results
Kickoff meeting Brief the team on the process and the expectations.		
Meeting #1 Business models and customer discovery What is	• All participants to develop a preliminary Business Model Canvas.	Concluded that BMC needed to account for existing business. Added blocks for current capabilities and

Workshop topic	Post-work	Results
marketing (4Ps). What are the challenges of SME B2B marketing? What is the Lean Start-Up? Why might it be applicable to SME firms. What is a Business Model Canvas (see Figure 1)?	• Develop a list of contacts for Customer Discovery	existing competition (see Figure 2).
Meeting #2 Value Proposition • Discuss customer pains and gains • Hypothesized pains and gains • The MVP • Common value proposition errors	• Scheduled meetings with two major customers. • Bring idea to two trade events (Phila and Las Vegas) • Developed script for all interactions	• Simplified customer segment • Added potential Value Proposition
Meeting #3	• Created strawman Survey	• Need to follow-up with

Workshop topic	Post-work	Results
Customer Segmentation • Start with Customer Canvas • Generate hypotheses • Rank pains and gains • Market types and Life Sales Curves	Monkey – facility size, number of automation needs, type and capacity of need. • Survey did not address hypothesis regarding productivity, call-out reliability, throughput, safety, Industry 4.0. • Talk to Engineering about design and documentation (user manual, how-to videos, etc.)	survey, only one response • Sales to reach out to 20 to 40 people to test hypotheses. • Put prototype on YouTube video • Send another survey request – sales will make follow-up calls with non-responders • Gathered F2F insights (Small machine shops, *Validated price!!,*

Workshop topic	Post-work	Results
		timing of production lines) • Three months for commercial design • Six more months for user documents
Meeting #4 Distribution Channels • Distribution as function of market complexity vs. solution complexity • Channel economics • Have we gathered enough reactions to MVP to test our hypotheses	• Tabulation of discussions to date suggests small manufacturers as niche	Feedback from customer visits include concerns about charging time for mobile system, stop-point accuracy, payload, speed, protection from other human operated equipment in the space.

Workshop topic	**Post-work**	**Results**
Meeting #5 Customer Relations		Hypothesis to test: SME manufacturing firms will use this product to increase productivity.
Meeting #6 Revenue Streams Review different revenue models	• Revenue models – direct sales – leasing to reduce entrance barriers – community of users	• Discuss with leasing companies as partners • Model and test time to close a sale.
Meeting #7 Resources, Activities, and Costs Resources: financial, physical, IP, and human	Resources, Activities, Costs	Estimate cash burn rate and time to cash positive
Meeting #8 Key Partners Suppliers, Lease Partners	Partners identified	

OUTCOME:

The project successfully developed and launched the automation project based on the Business Model Canvas of Figures 3 and 4. Detailed action plan that emerged from the workshop is listed in Figure 5. Market acceptance was robust enough that subsequent to a merger, the new management split the automation project off as a separate company that was more easily invested in.

Figure 1: Business Model Canvas as proposed by Alexander Osterwalder

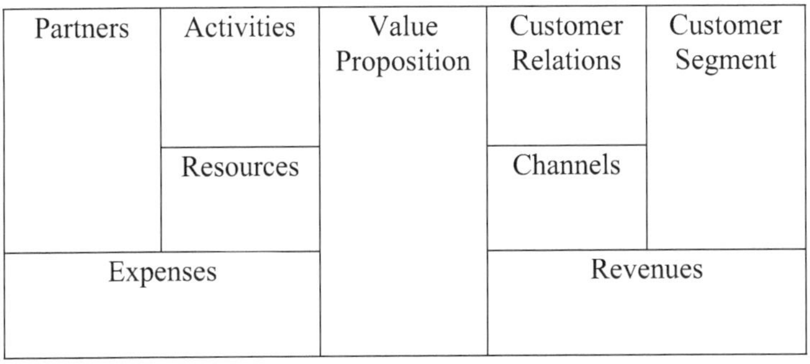

Partners	Activities	Value Proposition	Customer Relations	Customer Segment
	Resources		Channels	
Expenses			Revenues	

Figure 2: Business Model Canvas modified for existing businesses

<table>
<tr><td rowspan="3">Core
Capa
bility</td><td rowspan="2">Partner
s</td><td>Activities</td><td rowspan="3">Value
Prop</td><td>Customer
Relations</td><td rowspan="2">Customer
Segment</td><td rowspan="3">Com
petit
ion</td></tr>
<tr><td>Resources</td><td>Channels</td></tr>
<tr><td colspan="2">Expenses</td><td colspan="2">Revenues</td></tr>
</table>

Figure 3: Business Model Canvas of Automation Product Development

<table>
<tr><td rowspan="2">PARTNE
R

Suppliers

Integrators

Field
service

Leasing</td><td>ACTIVITY

Fabricate

Terms and
conditions

install</td><td rowspan="3">VALUE
PROP

Eff'y

Easy to
program</td><td>CUSTOMER
RELATION
Trade shows

Direct sales

Videos

Web ads</td><td rowspan="2">CUSTOM
ER
SEGMEN
T

SME

Mfr

Skid
transport</td></tr>
<tr><td>RESOURC
E
Marketing
and sales
consultant</td><td>CHANNEL

Direct sales

System
integrators</td></tr>
<tr><td colspan="2">EXPENSES
- BOM and assembly $L
- Installation
- Maintenance and support (need timeline)
- Spare parts
Tooling</td><td colspan="2">REVENUES
- Asset sale
- Lease
- Software is perpetual lease, not sale
-</td></tr>
</table>

Figure 4: Business Model Canvas – Post It notes indicate the amount of required iteration!

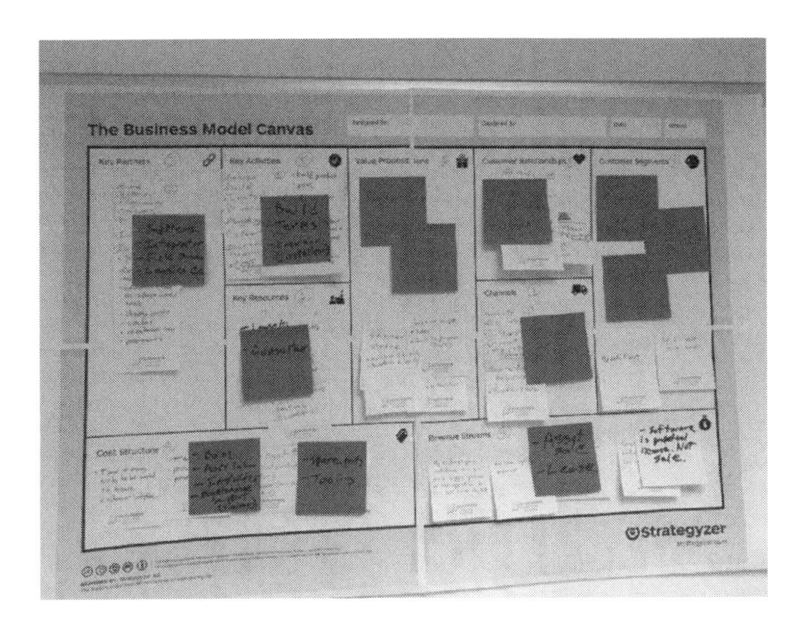

Figure 5: Lean Startup Workshop Action Plan

CRITICAL project MILESTONES

- Near term
  - 05 Jan – design requirements finalized
  - 15 Jan – identify chassis vendor
  - 15 Feb – all components on order
  - 08 March – chassis received
  - 05 March – all components received
- Longer term

- Terms and conditions for sale including software as perpetual license
- when do we need to identify partners for lease if asset sales arc not working?
- identify required stock of spare parts
- Identify out of area maintenance support
- Identify marketing consultant
- Generate sales department budget and manpower required to generate the required revenue by direct sales
- Engineering to develop software configuration management system
- Start discussions with indirect channel partners
- Identify key partners other than suppliers
    - Integrators, field service, and leasing company
- Product installations start with our engineering but this is not scalable
- Engineering to develop the maintenance support timeline

Bibliography

Osterwalder, Alexander et al. "The business model ontology: A proposition in a design science approach." (2004).

Chapter 5 – Architectural Innovation and competition against larger rivals

In an earlier chapter, we discussed the deep niche theory of SME innovation argued for by Gomes-Casseres in 1997. This theory argues that SME firms can successfully compete in deep niche markets where the SME firm is larger than the other SME firms in that niche. This is only helpful if an SME firm is in such a market. This model of SME innovation is of limited utility since SME management cannot control the nature of the market they compete in. This chapter argues for another innovation model that can empower an SME firm to compete against a larger rival, the model of Architectural Innovation.

One of the first projects that I evaluated and supported in my role as Corporate New Product Director (see *Chapter 2 – Using Case Study Research to Investigate SME Innovation*) was an innovative control system for a fan product that was a very successful and profitable product line for one of the corporate business units. The product idea was approved by our screening committee and the development effort proceeded successfully and the product was launched.

Within one year the control system was being ordered on 24% of the fan products sold by our business unit to industrial customers. The value of the control system option was not only the price of the option itself but the fan projects that we won due to this innovation and the price concessions that we avoided due to our unique capability. Needless to say, we were pleased with this project. But we were worried also. We were confident that our product was unique and this was eventually validated by the award of a utility patent by the USPTO. But we also knew based on our patent search work that there were ways around our patent that we could not protect against. We were constantly waiting for our larger rivals to counterattack and destroy our competitive advantage. But after six years (when I left the company) there had still not been a successful counterattack. There were products introduced to the marketplace to counter our product, but they never directly attacked the heart of our value proposition. Why not?

Recall that I was simultaneously running the corporate new product development effort and pursing doctoral studies in Innovation. I found the answer to "why not" in the 1990 work of Henderson & Clark in what they termed Architectural innovation. Their work refers to innovations based on the modification of product architecture. In his MIT doctoral thesis design expert Karl Ulrich defined product architecture as the linkages between product components that transfer the flow of mass, energy, load, and information. The theory of Architectural Innovation offers the insight that a significant competitive advantage can be gained from innovations that change only the manner in which the components work together as opposed to changing the components themselves. For instance, the typical drip-style coffeemaker consists of housing, filter basket, paper filter, ground coffee, carafe, power supply, heating element, water reservoir, and water pump. The highly successful Keurig coffeemaker was an architectural innovation in that it changed the linkages between the filter basket, paper filter, ground coffee and carafe.

Product architecture is often mirrored in the technical skills and managerial procedures of the firms producing the product. Firms manufacturing coffeemakers may have separate departments skilled at the molding of housings, producing filter baskets and carafes, power supply design, heating elements, and pumping water to a heating chamber. Firms would also develop procedures and problem-solving routines so the departments could collaborate. These skills and procedures become the firm's core capabilities, positioning it to exploit incremental component innovations and react effectively when competitors introduce incremental component innovations.

However, firms may not react well when faced with architectural innovations. As Dorothy Leonard-Barton discussed in her 1992 work *Core capabilities and core rigidities: A paradox in managing new product development* the core capabilities discussed in the previous paragraph can become core rigidities and inhibit the firm from effectively reacting to innovations offered by competitors, especially architectural innovations. The warning signs of architectural innovations may not be recognized due to the very skills and procedures built into the organization around the original product architecture. For instance, the introduction of the Keurig coffeemaker that uses a packet containing ground coffee with an integral filter to produce one cup would eliminate the need for a carafe and filter-basket. Firms with carafe and filter-basket departments might experience internal resistance in responding to this innovation. Their core capabilities have become core rigidities that inhibit their response to innovation. Firms that excel at building component core capabilities are often trapped in their original product architecture and suffer competitive failure when the market-accepted architecture shifts.

Architectural innovations have both marketing and technology implications and are often introduced by firms challenging the dominant firm in an industry. The dominant firm is often inhibited from introducing innovative architectures by the lack of appropriate skills and procedures or by being bound to what they know – the existing "successful" mindset. Additionally, the initial architectural innovations are often inferior to incumbent architectures when measured by parameters valued by customers served by the dominant firm. The low value placed on the initial performance of the architectural innovation further inhibits dominant firms from exploiting the innovation. The dominant firm risks a myopic view by being wedded to a specific product and its structure rather than providing value to a broad range of customers.

Challenger firms may introduce architectural innovations into adjacent markets considered unimportant by the dominant firm, but where the innovation has an advantage valued by the market. As the challenger deploys the architectural innovation into the adjacent market, they develop the expertise and cash flows to steadily improve the new architecture until it is superior to the originally dominant architecture. The dominant firm finds its market position challenged by a superior performing product architecture and could rapidly lose its dominant position to the new entrant with an architectural innovation.

In the case study under discussion, I believe this is what happened. The fan control system was an architectural innovation and our larger rivals failed to fully understand it. Additionally, we were focused on the adjacent niche market of laboratory exhaust fans. Our larger rivals were focused on supplying every fan in an industrial project. They did not recognize the nature of the innovation and were unwilling (or unable) to deploy the engineering resources required to effectively counter this architectural innovation in a way that was meaningful to the adjacent niche market.

SME firms can use the concept of Architectural Innovation by first explicitly understanding their product as an architecture and look for competitive advantages in innovating the linkages in the architecture. A solid strategy is to match these competitive advantages to adjacent markets that larger rivals are unwilling or unable to defend.

References

Gomes-Casseres, B. (1997). Alliance strategies of small firms. *Small Business Economics*, *9*(1), 33-44.

Henderson, R. M., & Clark, K. B. (1990). Architectural innovation: The reconfiguration of existing product technologies and the failure of established firms. *Administrative science quarterly*, 9-30.

Leonard-Barton, D. (1992). Core capabilities and core rigidities: A paradox in managing new product development. *Strategic management journal*, *13*(S1), 111-125.

Ulrich, K. (1995). The role of product architecture in the manufacturing firm. *Research policy*, *24*(3), 419-440.

Chapter 6 – Leading the New Product Development Effort – the Kotter Matrix

In 1990, John Kotter, a highly respected management scholar, published an article in the Harvard Business Review entitled *What Leaders Really Do* (Kotter, 1990). In this article, Professor Kotter makes a very practical argument for treating leadership and management as separate, yet integrated systems of action. He argues not for the primacy of one over the other, but for combining strong leadership and strong management. He begins his argument by first defining management as a system of actions we take to deal with the complexity of modern organizations, such as budgets, schedules, plans, and all the nuts and bolts that make an organization work. Leadership, on the other hand, is defined as the actions we take to deal with change, and getting people to understand that change might be necessary to effectively respond to turbulent environments, which are all too common in our organizations.

Professor Kotter further argues that each system of action (management and leadership) has three similar components. First, we must decide what needs to be done. Secondly, we need to create a mechanism to accomplish this goal, and finally we need to close the loop and ensure that people are taking the required actions to achieve the goal. This book applies the principles of Professor Kotter to new product development.

Deciding what needs to be done in the management system of action is the very familiar process of planning and budgeting. A plan needs to be established that includes a schedule, a budget, the required manpower, and all the details required to make something happen. Deciding what needs to be done in the leadership system of action is about establishing a vision. What new product capabilities will this organization require in five, ten or twenty years? What revered legacy products need to be improved, replaced, or even discontinued? And finally, what action needs to be taken right now to achieve the vision?

The second component of Kotter's model is to establish a mechanism for accomplishing the aforementioned goal. In the management system of action, this is about organizing and staffing. How are the new product teams organized? What reporting relationships are established? What problem-solving routines are required to work effectively with the rest of the business? Next, you face the always-challenging task of putting the correct people in the correct jobs. On the other hand, the leadership system of action calls on us to align people to the vision we have established. This is more of a communications challenge than the management challenge of organizing and staffing. We must effectively communicate the new direction our vision requires, and create coalitions that not only understand this vision, but will commit to achieving it.

The final component of the Kotter model is what I refer to as closing the loop, or how to ensure that the work gets done! In the management model, this is about controlling and problem solving. How is progress monitored to detect variances so that corrective action can be taken? Will you employ a stage-gate system, will you employ design reviews, will you design a “qualification test” to prove that the new product is ready to be deployed? In the leadership system of action, motivation and inspiration is needed to encourage people to pursue the established vision. This is critical since change is always difficult, and people need to feel inspired to overcome the unanticipated barriers they are sure to encounter.

I have always found John Kotter's model extraordinarily useful in diagnosing organizational challenges. I rely on a visual form of it that I refer to as the Kotter matrix, depicted in Figure 6. If you have read Kotter's paper, and I encourage you to do so, you will note that he did not organize his thinking into this visual model. I wish I could claim authorship of organizing these constructs into a visual form, but it was introduced to me by Greg Bruce, emeritus dean of the business school at La Salle University in Philadelphia. Not only have I found it useful, but throughout the years, my students have found it to be a very helpful tool in the diagnosis of organizational problems, including new product development.

If you examine Figure 6 you can see how it can be used to organize your thinking. The left-hand column is the entire management system of action, while the right-hand column is the entire leadership system of action. The rows include each of the three actions that each system needs to accomplish: deciding what needs to be done, the mechanism for doing it, and how to close the loop and make sure the work really gets done.

The visual model of the Kotter matrix has the added value of demonstrating how leadership and management interact; why it is important to be good at both; and why one is not an adequate substitute for the other. For example, examine the first row of actions, which include budget and plan under the management column, and vision under the leadership column. Let's assume that you spent a lot of time and energy trying to understand your organization, the outside environment, and how that environment is changing. All this effort resulted in the development of a useful and valid new product vision for your organization. This is terrific, but what's next? Your vision will not become a reality unless you move from the leadership column to the management column. You will need to pivot and put plans and budgets in place to start making things happen right now in order to achieve your vision, otherwise your chances of accomplishing it are slim. The vision must be turned into action. This is the first argument supporting Kotter's contention that leadership and management may be distinct systems of action, but they are highly complementary—you need both.

Once you have your budget and plan, the next obvious step is to move down the management column to organize and staff. What organizational design or existing departments will carry out your plans, and who will do the work? When you have accomplished this step, you need to make sure all employees understand and believe in the vision, and are willing to work towards it. In other words, you need to move back over to the leadership column and get everyone aligned to the vision. Once this has been accomplished, you can think about the bottom row of Figure 6 and put both problem detection and control routines in place (the management column), as well as decide what you need to do to motivate and inspire the team to pursue the vision, despite the obstacles they will undoubtedly encounter (the leadership column).

The Kotter matrix is an extremely valuable tool that can help you understand how your leadership and management actions interact, and how that interaction can make your organization more effective.

The Kotter matrix can also be used to diagnose new product development problems. It is always an effective first step in unraveling what's going wrong, so that action can be taken. For instance, if an organization is very good at the management column of the Kotter matrix, but not so good at the leadership column—a common story in my classes—then this helps diagnose where you need to look for a solution. If you're not getting the new product results you want because your vision is disconnected from environmental realities, then working harder on the management side of the matrix by budgeting better or controlling and problem solving better is not going to fix the problem. The problem is that the organizational vision, whether stated or unstated, is not aligned to the environment and you need to get it into alignment. Likewise, if your budgets are unrealistic, you can do all the visioning and aligning your heart desires, but you will not fix the problem. The challenge is to resolve the disconnect between the resources you have devoted to a new product development project and the resources required to get the outcome you desire. So, the Kotter matrix is a great first step in the diagnosis of most organizational problems, and it serves as a great framework for planning new initiatives. The Kotter matrix helps us identify where problems are and where action needs to be taken.

Figure 6

The Kotter Matrix

MANAGEMENT (Complexity, Things)	***LEADERSHIP*** (Change, People)
Budget & Plan	Vision
Organize and Staff	Aligning People ("buy in" to vision, networks)
Control & Problem Solve	Motivation and Inspiration

In my graduate management classes I ask students to bring in "case stories" that they have lived through or witnessed first-hand. We then apply class principles to analyze these real-world situations and determine if different actions could have resulted in better outcomes. I believe that this makes a stronger connection between management theory and real-world applications. This is one such "case story"

The Start-up Case Story

A student shared this experience of a small entrepreneurial start-up company working on an innovative product that would use information derived from genetic testing to increase the effectiveness of medical treatment. Since this company was a start-up without a lot of funding, employees were incentivized by equity in the company that would pay off in the future. The company's sales representatives would go to doctors' offices, demonstrate the service, and sign the doctor up as a customer.

The CEO recruited employees by promising that in a few months the company would be worth millions of dollars and that then, the team would have little to no work to do. The CEO had no sales experience and refused to go out with the sales team, citing that he was too busy with managerial duties. Also, the target market he selected imposed 4.5- to 6.5-hour daily commute on each sales representative. The sales team used their own cars and money for gas and tolls. At first, they were enthusiastic because the CEO was promising that they would benefit from this work in a few short months.

After a few weeks of long hours and tiring commutes, the sales team found that it was more difficult to sign up doctors for the service than had been anticipated. However, the CEO rejected the idea that the business model could be flawed, and accused the sales team of failing. He would single employees out and tell them how they were doing everything wrong, yet he refused to travel with the sales team due to the lengthy commute. The team's enthusiasm for the mission began to fade away because of the extensive travel and having to spend their own money without reimbursement.

Finally, after about a month, one sales representative convinced a doctor to sign on to the service. The CEO was thrilled, though instead of celebrating the sale as "a team win," he used the occasion to criticize every other team member who had yet to make a sale. He told them they were slacking, and they were easily replaceable. He praised the sales representative who had closed the first sale, and publicly stated she was the only one putting in hard work, which, of course, made the other sales team members angry. As a result, the other team members started to show a negative attitude towards her and eventually the group did not function as a team anymore. However, the CEO did not care; he just wanted to drive sales. Most of the sales representatives ended up leaving the company because it was too much work with no reward, and there was constant pressure from the CEO. The start-up subsequently failed.

In this situation, there are many things that were done poorly and could have been handled better. But, let's stick with what we can learn by the application of the Kotter matrix. The CEO clearly had a vision of growing a company based on the value of genetic testing services for doctors. He clearly inspired his team with equity and the promise of future wealth. If we assume that the technology worked and that the value proposition was reasonable, then this CEO followed the leadership side of the Kotter matrix well. He developed a vision, aligned his team to that vision and inspired them initially. However, the management side of the Kotter matrix was not handled with the same skill. The vision needed to be turned into plans and budgets. The plans were obviously problematic, in that the sales cycle was longer and more difficult than had been anticipated. The CEO would have been well-served to travel into the field with his sales representatives and see exactly where his plan was insufficient. Then, a new plan with the appropriate budget could have been put in place. Let's assume that the staffing was appropriate and that this company had the correct sales representatives. Is a company really organized correctly when all of its sales representatives have a lengthy commute to their target market? These should have been questions from the outset. The final part of the management side of the Kotter matrix is controlling and problem solving. Sales were not being made, so detecting the problem was fairly straightforward. A more effective way to solve the problem could have been used rather than criticizing the sales representatives and hoping that the unreasonably high expectations would spur them to overcome whatever obstacles they were faced with.

This case story is a good example of how we can use the Kotter matrix to focus on the problem area. The first issue I would look at is the plan for achieving the vision. Do we really know how long it takes to sell a doctor on this service or what information the doctor requires to make the purchase decision? Secondly, does it make sense to employ a sales force that's located six hours away from your target market? Once the plan and the associated budget is reviewed and improved, the CEO should go about executing the optimized plan. Obviously, this could lead to discovering the next barrier that must be overcome, but that's not necessarily a bad outcome. So, this is a good illustration of how using the Kotter matrix to think about an organizational situation can lead you to where the problem is. It doesn't necessarily tell us how to solve the problem, but it guides us towards where to look. I think it would be a mistake to look at this situation and assume that the vision of the CEO was flawed, and that the business doesn't have a chance. I make this conclusion since the management of the effort was so poorly executed that even a great vision would not have had a chance. If we were to encourage the CEO to revisit the vision and perhaps revise the vision, we still would probably not be successful because the same problematic plan would be used. In this instance, the management side of the Kotter matrix must be adjusted first.

Actions to Take

- Referencing Figure 6: Establish a clear vision and how the new product development effort will support that vision. A vision can be inspiring and altruistic, but it doesn't have to be. A vision of providing the best technology to the customers in your niche and providing a safe and inspiring workplace for employees is a great vision! The important thing is that you know what your vision is.
- Referencing Figure 6: Develop a plan and budget that can propel your new product vision into action. Having a vision of providing the best technology is

empty if you don't invest in developing the technology. The point is that your vision must be connected to budgets and plans if it is to mean anything.

- Referencing Figure 6: Get the right people in the right jobs so your company can successfully execute the plan.
- Referencing Figure 6: Socialize the vision and demonstrate why it's critical to team members so that they are able to buy in and commit to it. Assure that everyone is aligned.
- Referencing Figure 6: Put a system in place to detect problems so that you can address them effectively. This can be a straightforward quarterly review of the plans and budgets required to actualize the vision. It can include design reviews, a stage-gate system or a qualification test. When you put a plan in place to enact your vision you must measure it to see if you are making progress. If not, then identify the problem and solve it.
- Referencing Figure 6: Keep your team inspired and motivated.

References

1. Kotter, J.P. (1990). What Leaders Really Do, *Harvard Business Review, 68(3)*, 103 – 111.

Chapter 7 – Leading the New Product Development Effort – Transformational leadership, transactional leadership, and Stage-Gate

Management thinker Peter Drucker (1987) describes systemic innovation as "the purposeful and organized search for changes, and the systemic analysis of the opportunities such changes might offer for economic or social innovation". Drucker's call for systemic analysis was answered by the introduction of the now ubiquitous stage-gate process as a systemic methodology to develop new product innovations that will achieve technological, market and financial success (Cooper and Kleinschmidt, 1986). Since 1986 the stage-gate process has become so common that repeating the details is not called for here. We'll just characterize the stage-gate process has an application of rigorous analysis to an idea as it matures through successive stages of technological and marketing development stages and review gates. However, the original stage-gate methodology did not provide a thorough description of how to generate the initial idea for an innovative product. Subsequent modifications were suggested to better capture how a new product development process could develop the initial ideas, including an iterative method of customer analysis (Cooper, 2001), a front-end process that integrates "seemingly disparate but related strategic and operational activities, typically crossing functional boundaries" (Khurana and Rosenthal (1998), a "fuzzy front end" that iteratively evaluates technology and market opportunities, and others. Combining these "front end" processes with the stage-gate process provides a more complete understanding of the innovation development process.

However, empirical studies suggest that this approach is effective with only a sub-set of innovation types. First let's define two broad categories of innovation. Research scholars refer to innovations that exploit a firm's current knowledge as exploitive innovations while innovations that require the exploration of new knowledge are defined as explorative innovations. It should be noted that this definition is based on exploring knowledge that is *new to the firm*, not necessarily new to the world. A twenty-year study by Benner and Tushman (2002) concluded that the application of "process management activities" to new product development processes will increase the number of innovations that exploit a firm's existing knowledge base at the expense of innovations relying on the exploration of new competencies. Many aspects of stage-gate processes fit the description of the process management activities studied by Benner and Tushman (2002) leading to the conclusion that the rigorous application of process management activities at the heart of the stage-gate process will be predisposed to develop exploitive new products at the expense of explorative new products. Management research scholars Tushman & Smith in a 2002 study note that both exploitive and explorative innovations make unique contributions to a firm's competitiveness. They argue that an ambidextrous firm capable of developing both types of innovation will maintain a competitive advantage (Tushman& Smith, 2002).

The recognition by scholars that our current understanding of new product development processes lacks the ability to explain the emergence of both exploitive and explorative innovations coincides with calls for the use of multiple new product development processes within a single organization (MacCormack, Crandall, Henderson, & Toft, 2012) and practitioner calls for innovation processes that can react faster to the realities of some markets (Blank, 2013). This chapter presents an approach to providing organizations with such an ambidextrous innovation process to pursue both exploitive and explorative innovations.

A parallel stream of inquiry into the basis of successful innovation has addressed the role of leadership. Early thinking proposed that R&D "stars" would perform a gatekeeping role that translates valuable external information for use by internal development actors (Allen, 1977). The importance of an individual innovation advocate was expanded to include the role of the "heavyweight" project manager whose boundary spanning capabilities contributes to the success of innovation efforts (Fujimoto and Clark, 1991). Investigations into the impact of individual leadership on innovation were reignited when the differentiation between transformational and transactional leadership was introduced (Bass, 1985). The introduction of this dichotomy into the literature sparked a number of investigations into the link between transformational leadership and the success of innovation efforts. There is significant evidence in the literature that transactional leadership is more effective in structured situations while transformational leadership is a more effective approach for unstructured environments. Yet extant work does not explicitly consider how these leadership types influence the two-stage innovation process (the combination of a fuzzy front-end and a rigorous stages and gates process), or which leadership type will be most productive for organizational innovation.

This chapter proposes a conceptual framework to resolve the conflict between a process designed for the development of exploitive innovations and one that can accommodate the emergence of explorative innovations This ambidextrous innovation process is based on the recognition that all of the competing concepts for a new product development process can be distilled into a two-phase generalized model where the first phase generates ideas in a loosely defined and uncertain way (sometimes referred to as the fuzzy front-end) and the second phase develops and evaluates these ideas in a more prescribed and predictable stage-gate process.

In this chapter I argue that an ambidextrous product development process will require ambidextrous leadership and will apply the different strengths of both transactional and transformational leadership to suggest an effective leadership approach. The proposed conceptual framework will therefore combine the literature stream of innovation processes with the stream of innovation leadership by proposing that the turbulent portions of the process are best managed by a transformational approach while the predictable steps will yield better results when managed by a transactional approach. The proposed framework provides an integrated innovation development strategy that can be applied across a broad range of environmental and technological uncertainties.

Additionally, the proposed framework argues for the application of the appropriate leadership approach to different phases of the process.

What I am proposing is that how an innovation process is led is at least as important as the structural design of the process. Since the fuzzy front-end is turbulent and unpredictable, a transformational leadership style is better suited and more likely to result in explorative innovation ideas that can be then introduced into the stages and gates portion of a firm's innovation process. The more prescribed, predictable, and rigorous stage-gate process could then be led transactionally.

But how exactly does an innovation manager lead transformationally? Let's consider transformational leadership as articulated by Bass and Avolio (Bass, 1990). Bass and Avolio differentiate between transformational and transactional leadership. Transactional leadership is simply a transaction! Your team member performs the task required of them and you provide the reward that you promised them. If they sell the car, then you pay them the commission. If they sell the stock, then you pay them the commission. If they get the design done on time, then you tell them that they did a good job. It is strictly a transaction. Transformational leadership on the other hand is about elevating the interests of the staff members. It's about generating a sense of mission or an awareness of your mission in your team members. The thinking is that they will be more motivated and work harder to accomplish a mission that they have "bought into." Since Bass and Avolio first presented the theory, there have been many empirical investigations into it. This research concluded that you are more likely seen as an effective leader by your team members if you lead transformationally. Research has found higher organizational outcomes, including better financial performance, when teams are led transformationally. And there is the valuable finding that teams work harder for transformational leaders then they do for transactional leaders.

When I discuss this material with experienced leaders in a face-to-face setting, I will usually pose the question of which is better: transformational or transactional? It usually doesn't take very long for someone to note that transformational leadership sounds great, but sometimes you need to be transactional. In other words, you need to be able to do both. I agree with this and so do Bass and Avolio. Transformational leadership is only one tool in your leadership toolkit. There is a place for transactional leadership. It's also important for us to understand that if we look at transformational and transactional leadership as the endpoints of the spectrum, most of us will fit somewhere between these two extremes. Our center of gravity, if you will, may be towards the transformational end of this spectrum or perhaps towards the transactional end of this spectrum. Naturally, most of us will tend towards one end or the other, and very few of us will be good at both—although being good at both is certainly something to strive for! This self-awareness is important because if a situation requires transactional leadership and we are not very good at it, then we need to act to ensure the right steps are taken, such as stretching ourselves to take on tasks that may not be our strength or delegating.

Transactional leadership has been shown to work quite well in stable environments. Think of the previous example of selling cars. Leadership will be able to predict with confidence what actions are required for success and design reward systems (the transaction!) that encourage those actions. The business environment simply doesn't change that much each day; therefore, the transactional approach can work quite well. Transactional leadership can also be applied well to the management of stage-gate processes. Each gate requires specific deliverables (budgets, marketing estimates, etc.). They are either adequately done or not. They either predict the success or failure of the new product development effort. The path that a new product idea takes through the stage-gate process is quite predictable and transactional leadership can be effective.

Transformational leadership has been shown to work better in turbulent environments where it is more difficult to predict what the business environment will be like on a day-to-day basis. Not surprisingly, there is a large body of work correlating transformational leadership and successful innovation efforts. The application of transformational leadership in the idea generation phase of a new product development process will increase the likelihood of explorative new ideas. The subsequent application of transactional leadership will help the ideas move through the stage-gate phases.

So how do we do this thing called transformational leadership? Bass and Avolio suggest what they refer to as the 'four i's' as the foundation of transformational leadership. The first 'i' is **idealized influence**, which is sometimes referred to as charisma or vision. For our purposes, charisma

does not refer to what we normally associate with a Hollywood star. It refers to the ability to get people excited about a future state they can achieve by working towards it. It also refers to communicating your vision of this future state, and getting your team members to buy in and commit to it. The second 'i' is the **inspiration of high expectations**. People are inspired and motivated when their leadership sets challenging—but realistic—goals and supports their followers in pursuing these goals. "The third 'i' is **intellectual stimulation** or working with your team members to help solve challenging problems. And the fourth 'i' is **individualized consideration**, such as coaching and advising. These 'four i's' are repeated in Figure 7 for your reference. The 'four i's' give leaders a very practical method of
leading transformationally.

Figure 7: The "four i's" of transformational leadership

Transformational Leadership
Idealized influence (charisma or vision) Inspiration (high expectations) Intellectual stimulation (careful problem-solving) Individualized consideration (coaching, advising)

Case Story

Let's think back to Chapter 6 and the case story of the startup selling its innovative products directly to doctors' offices. In this story, the CEO recruited team members by convincing them that the company would soon be very successful and they would share in the financial success. He must have been quite convincing since the sales representatives were motivated enough to use their personal vehicles and money for commuting expenses to support the company's sales efforts. This would indicate that the CEO of this company successfully utilized the first 'i,' or the idealized influence of a vision.

However, let's follow the actions of this CEO when his team had a hard time making sales. He refused to accompany the sales team when they visited doctors' offices. Had he done so, he would have experienced the sales representatives' problems first hand, which then would have enabled him to engage in careful problem-solving with his team. This is the third 'i,' intellectual stimulation. Without this in-depth understanding of the challenges the sales representatives face, he was unable to coach or advise his team on how to improve their performance. In other words, he was unable to apply the fourth i, individual consideration. Also, in Chapter 6 we discussed that when the first sale was made, the CEO used this as an opportunity to berate the other sales representatives as easily-replaceable slackers. This first successful sale became a lost opportunity to understand how the sale had been made and to apply this knowledge to problem solving and coaching with the other sales representatives.

In this case story, the CEO could wield idealized influence, or vision, by making promises. However, when difficulties arose, the CEO failed. If this startup leader had fully understood the concepts of transformational leadership, he might have applied intellectual stimulation and individualized consideration, and therefore could have perhaps enjoyed a more successful outcome.

The unfortunate demise of this startup company with an innovative new product can be directly attributed to an incomplete understanding and application of transformational leadership. Much is written about the power of a vision, but the work of Bass and Avolio argues that a vision is only the starting point of transformational leadership. We need to be skilled at all aspects of transformational leadership— all four of the 'i's.'

Actions to Take

These are the specific actions that can be taken away from this chapter to apply transformational leadership to the management of the idea generation phase of your new product development effort.

- Reflect on your business environment. Is it more turbulent or stable? Transformational leadership has been shown to work better in turbulent environments where it is more difficult to predict what the environment will be like on a day-to-day basis.
- If your approach should be more transformational, determine if you are taking the following actions:
    - Do you have a vision for the business? How does new product development support this vision? Does this vision make sense to the major stakeholders? This is the first 'i,' or idealized influence, of transformational leadership.
    - Are you communicating your vision for the business and getting your team members to buy in and commit to it? Do not underestimate this task. This communication is not a one-time event. You must constantly communicate the vision.
    - Are you inspiring your team with high expectations? This is the second 'i.' A word of caution here is that the expectations must be realistic. Unrealistic expectations have been proven to damage motivation.

- Do you work with your team members to *help* solve challenging problems? This is the third 'i,' or intellectual stimulation.
- Do you coach and advise your team? This is the fourth 'i' or individualized consideration.

References

Allen, T. J. 1977. ***Managing the flow of technology: Technology transfer and the dissemination of technological information within the R & D organization***. *Research supported by the National Science Foundation. Cambridge, Mass.; MIT Press, 1977. 329 p.*

Bass, B. M. (1985). Leadership and performance beyond expectations. New York: Free Press

Bass, B. M. (1990). From transactional to transformational leadership: Learning to share the vision. *Organizational dynamics*, *18*(3), 19-31.

Benner, M. J., & Tushman, M. 2002. Process management and technological innovation: A longitudinal study of the photography and paint industries. ***Administrative Science Quarterly***, 47(4), 676-707.

Blank, Steve. 2103. "Why the lean start-up changes everything." ***HARVARD BUSINESS REVIEW*** 91.5: 64-+.

Cooper, R. G. 2001. ***Winning at new products***. London: Kogan Page.

Cooper, R. G., & Kleinschmidt, E. J. 1986. An investigation into the new product process: steps, deficiencies, and impact. ***Journal of product Innovation management***, 3(2), 71-85.

Drucker, P. F. 1987. ***Innovation and entrepreneurship***. Newbridge Communications.

Fujimoto, T., & Clark, K. 1991. ***Product development performance. Strategy, Organisation and Management in the World Auto Industry***. Cambridge: Harvard Business Press

Khurana, A., & Rosenthal, S. R. 1998. Towards holistic "front ends" in new product development. ***Journal of Product Innovation Management***, 15(1), 57-74

MacCormack, A., Crandall, W., Henderson, P., & Toft, P. 2012. Do you need a new product-development strategy? Aligning process with context. ***Research-Technology Management***, 55(1), 34-43.

Tushman, M. L., & Smith, W. 2002. Organizational Technology. In J. A. Baum (Ed.), ***The Blackwell companion to organizations*** (pp386 - 414). Blackwell Publishers.

Chapter 8 – The New Product Development Plan

In this final chapter I will refer to conclusions that we reached in previous chapters and outline a plan for new product development for SME firms.

ACTION 1: Establish a vision of how new product development will contribute to your business. A discussion of how to develop an effective vision is beyond the scope of this book. Effective visions are based on mental models of how your market works. These models are built by gathering information from a wide variety of stake holders. A vision doesn't have to be inspiring and altruistic, but it needs to serve the needs of multiple stakeholders and it needs to be achievable[2].

ACTION 2: Establish an NPD budget. It is important that the budget is realistically aligned with the vision that you articulated in Action 1. If not, the vision and the budget need to be brought into alignment. The focal firm in the case study (Chapter 2) achieved meaningful results with a budget of 1% of revenue[1,2].

ACTION 3: Establish an NPD team that is cross disciplinary and not strictly technical. The specific skills on this team (and the amount of time they can commit) must be capable of executing a plan that can achieve the new product vision articulated in ACTION 1 utilizing the budget established in ACTION 2. This team needs to evaluate both the business case for proposed innovations and the technical feasibility[1,2].

ACTION 4: Establish a new product portfolio. New product development is inherently risky and failures will occur. It is important to manage the risk by developing a portfolio of NPD projects that are both incremental and disruptive. Risk can further be managed by eliminating weak projects as soon as the weakness becomes apparent. Embrace the concept of FFFC (Fail-Fast-and-Fail-Cheap)[1].

ACTION 5: Establish a formal new product development process. The NPD team or office must articulate the NPD process. The process should be formally documented and followed! At regular intervals the NPD results should be evaluated and the process modified as necessary. The NPD team should consider Stage-Gate, Design Thinking, and Customer Discovery processes. Take the time to understand if transformational leadership is required at any step in the new product process. This is most likely to be necessary in ideation and creative problem-solving efforts. Is your team leadership capable of leading transformationally when required. If not, training or additional talent may be required. [1,2,3].

ACTION 6: The SME firm should invest the time to understand their products as an architecture of managing load, energy, mass, and information. Also understand the architecture of the products of the firm's key rivals in the marketplace. Now the firms can be creative in applying architectural innovation and searching for adjacent markets[1,4].

ACTION 7: Develop a formal NPD marketing effort. The three primary NPD marketing efforts are new product ideation, market research to support a Stage-Gate process, and product launch. These topics are thoroughly covered elsewhere and they don't bear repeating here. I will only add two caveats. First, how to sell a product should be pilot tested with the same rigor as developing the product. Secondly, when launching a new product, pay attention to the *entire* distribution chain including your internal personnel who specify products in response to customer inquiries. How you sell the new product should add value to every step in the sales channel[1].

ACTION 8: Determine if your market fits the definition of a deep niche market (a number of similar sized firms selling business-to-business products). *Rigorously* determine if you are the technology leader as understood by customers. Then generate a technology development plan to either retain or obtain technological leadership in your deep niche[1].

1. *From the previously unpublished theoretical model for SME firm innovation (Chapter 3)*
2. *From Chapter 6 – Leading the New Product Development Effort – the Kotter Matrix*
3. *From Chapter 7 – Leading the New Product Development Effort – Transformational leadership, transactional leadership, and Stage-Gate*
4. *From Chapter 5 – Architectural Innovation and competition against larger rivals*

Made in the USA
Middletown, DE
15 August 2024